AGAPANTHUS GARDENING HORTICULTURISTS GUIDE FROM CULTIVATION TILL COMMMERCIAL SUCCESS

Expert Cultivation Techniques, Care Tips, And Commercial Success Strategies For Horticulturists

MANUEL SHELTON

Copyright © 2024 by Manuel Shelton

All rights reserved. No part of this publication may be reproduced, distributed, or transmitted in any form or by any means, including photocopying, recording, or other electronic or mechanical methods, without the prior written permission of the publisher, except in the case of brief quotations embodied in critical reviews and certain other noncommercial uses permitted by copyright law.

Disclaimer

The views expressed in this book are solely those of the author and do not necessarily reflect the views of any company, organization, or individual.

The author is not engaged in any endorsement deals or partnerships with any entities mentioned in this book. Any references to products, services, or organizations are for informational purposes only and do not constitute endorsement.

Readers are encouraged to conduct their own research and exercise their own judgment before making any decisions based on the information provided in this book."

Contents

CHAPTER ONE 13
COMPREHENDING AGAPANTHUS'S ORIGINS AND VARIATIONS 13
- Advantages Of Growing Agapanthus 14
- Creating Objectives For Your Agapanthus Landscape 16

CHAPTER TWO 19
SETTLING UP 19
- How To Choose The Best Site For Your Agapanthus Garden 19
- Getting The Soil Ready For Maximum Growth 20
- Selecting The Appropriate Agapanthus Types For Your Climate 22

CHAPTER THREE 25
AGAPANTHUS PLANTING 25
- A Comprehensive Guide For Planting Agapanthus Seeds Or Bulbs 25
- Advice On Fertilisation And Watering For Healthful Growth 27
- Mulching Strategies To Preserve Weeds And Preserve Moisture 28

CHAPTER FOUR 31
  TAKING CARE OF AGAPENTHUS 31
    Essential Agapanthus Care Practices 31
    Techniques For Controlling Pests And Diseases In Agapanthus 34
    Understanding Stress Symptoms And How To Treat Them 35
CHAPTER FIVE 39
  DISSEMBLING AGAPANTHUS 39
    Methods Of Propagation 39
    Strategies And Timing For Effective Propagation . 43
    Solving Typical Propagation Problems 44
CHAPTER SIX 47
  UTILISING AGAPANTHUS IN DESIGN 47
    Using Agapanthus In Landscape Architecture 47
    Ideas For Stunning Combinations Of Companion Plants 48
    Advice For Designing Agapanthus Arrangements To Make A Visual Statement 49
CHAPTER SEVEN 53
  AGAPANTHUS WITHIN CANDS 53

Selecting The Proper Soil Mix And Containers For Potted Agapanthus 53

Tips For Caring For Containers: Watering And Fertilising 54

CHAPTER EIGHT 59

A AGAPANTHUS FOR SUCCESS IN COMMERCE 59

Prospects For Agapanthus Growers In The Market 59

Increasing Production To A Commercial Level 61

Sales And Marketing Techniques For Agapanthus Products 63

CHAPTER NINE 65

FAQs AND REGULAR QUESTIONS 65

Responses To Common Questions Concerning The Cultivation Of Agapanthus 65

Taking Care Of Typical Problems Like Yellowing Leaves And Insufficient Blooms 67

A Beginners' Guide To Troubleshooting 68

CHAPTER TEN 71

PROGRESSIVE METHODS AND INNOVATIONS 71

Sophisticated Cultivation Methods For Skilled Farmers 71

Advances In Agapanthus Research And Breeding 73

Examining Novel Approaches And Opportunities In Agapanthus Horticulture 75

CONCERNING THIS BOOK

"Agapanthus Gardening with Elegance" is more than simply a book; it's a thorough guide that provides access to a world of exquisite flowers and skillful horticulture. This book, which explores the essence of Agapanthus, starts with a perceptive look at its history and many variants. It highlights the many advantages of growing agapanthus, from its eye-catching blossoms to its adaptability to a wide range of climates. With this introduction, readers will set out on a path that goes beyond simple gardening—a quest to develop grace and refinement in their outdoor environments.

Readers receive priceless advice to get them started in gardening as they explore the practical side of agapanthus growing. Every step is carefully laid out to guarantee ideal development and vibrancy, from choosing the ideal spot to precisely preparing the soil. The book acts as a beacon, pointing enthusiasts in the direction of the best Agapanthus cultivars suited to their particular climate. Equipped with this understanding, they confidently start the planting process by adhering to an easy-to-follow, step-by-step manual that is full of advice on mulching, fertilization, and watering methods.

The careful instructions in this book turn caring for Agapanthus into an artistic endeavor. From watering schedules to trimming strategies, essential routines are demystified, enabling readers to perfectly care for their plants. In addition, the book provides a thorough review of disease and pest management, guaranteeing that any obstacles faced are promptly and accurately resolved.

As they turn each page, readers gain proficiency in identifying stress indicators and rehabilitation techniques, resulting in a positive relationship between them and their plant friends.

The book explores the topics of propagation, design, and economic feasibility in addition to cultivation. The author encourages readers to let their imaginations run wild and incorporate Agapanthus artfully into their landscapes by providing thorough descriptions of propagation techniques and creative design ideas. Furthermore, knowledge of container gardening and business potential turns hobbyists into entrepreneurs, enabling them to profit from their botanical expertise in addition to beautifying their environment.

Readers become experts as they work their way through FAQs, troubleshooting manuals, and advanced techniques, and they take an even greater interest in Agapanthus. The book turns into a reliable companion, revealing the most recent advancements in Agapanthus breeding and research while providing answers to frequently asked questions.

In the end, "Agapanthus: Gardening with Elegance" is more than just a gardening guide; it's a monument to the creativity and inventiveness that arise when enthusiasm and understanding combine.

Overview of Agapanthus

Agapanthus, sometimes referred to as African lilies or Lily of the Nile, is a gorgeous flowering plant that brings beauty and elegance to any garden setting. This perennial beauty, which is native to South Africa and is a member of the Amaryllidaceae family, is prized for its eye-catching globe-shaped clusters of blue or white flowers perched on tall stalks. Its thick, strap-like leaf adds even more visual appeal, which is why gardeners all across the world choose it.

CHAPTER ONE

COMPREHENDING AGAPANTHUS'S ORIGINS AND VARIATIONS

To properly cultivate Agapanthus, one must be aware of its history and the range of available cultivars. Agapanthus is a native of southern Africa and prefers conditions with warm summers and moderate winters. But it may also be cultivated in colder climates with the right selection and maintenance.

There are many different species and cultivars of agapanthus, and each has its special qualities regarding bloom color, size, and growth habit. Agapanthus africanus, Agapanthus praecox, and Agapanthus orientalis are the most prevalent varieties. Praecox variants bloom earlier in the season, although Africanus is recognized for its hardiness and usually has blue flowers. Orientalis cultivars might have larger flower heads and exhibit white, blue, or purple hues.

Take into account your climate, the type of soil in your yard, and the amount of space you have while choosing Agapanthus. To guarantee ideal growth and bloom, select cultivars that are compatible with the environmental requirements of your area.

Advantages Of Growing Agapanthus

There are several advantages to growing agapanthus besides its visual attractiveness. Here are some strong arguments for adding this lovely plant to your garden:

Low upkeep: Once established, agapanthus requires very little upkeep, making it a reasonably easy plant to care for. Being tolerant of drought and thriving in diverse soil types, it's a great option for individuals who are busy with gardening or have little time for maintenance.

Long Blooming Period: Agapanthus can bloom for a long time, usually from late spring to early autumn, if they receive the right care. Its profusion of blooms draws pollinators like bees and butterflies to the garden while also adding color and aesthetic intrigue.

Versatile Use: Agapanthus works well in a variety of landscape styles, whether it is used as a focal point in garden beds, borders, or containers.

It works well in both formal and casual garden settings thanks to its upright growth habit and eye-catching blossoms.

Erosion Control: Agapanthus is a great option for controlling erosion on hillsides or slopes because of its strong root system, which stabilizes the soil. Difficult environments can be made more beautiful and less prone to soil erosion by planting agapanthus along banks or terraced areas.

Potential for Cut Flowers: Agapanthus's long, robust stems make it perfect for setups involving cut flowers. Gathering blooms for indoor display brings their beauty closer to you and elevates the look of your interior design.

Creating Objectives For Your Agapanthus Landscape

To achieve success and to direct your efforts, it's important to create specific goals before beginning an Agapanthus gardening project. Take into account the following elements while designing your garden with agapanthus:

Design Aesthetic: Choose the general style and mood you wish to accomplish for your Agapanthus landscape. Plant arrangement and selection should be in line with your preferred style, whether that is a more casual, naturalistic design or a formal, structured layout.

Flower Colour and Variety: Select Agapanthus cultivars that blend in with your current garden's color scheme or that will bring a splash of color to particular spots.

To add visual appeal and balance to the landscape, think about combining various cultivars and colors.

Maintenance Level: Choose Agapanthus types that correspond to the degree of upkeep you wish to achieve by evaluating the time and resources you have available for garden maintenance.

If you don't have much time for watering, fertilizing, and trimming, go for low-maintenance kinds.

Seasonal Interest: Include more plants that go well with Agapanthus and add interest at different times of year to create year-round interest.

To create a dynamic and lively garden landscape, incorporate decorative grasses, seasonal bloomers, and evergreen foliage.

Environmental Considerations: When designing your Agapanthus garden, consider elements including soil type, climate, and sun exposure. To guarantee wholesome growth and lifespan, pick kinds that flourish in the particular environmental circumstances that you have.

You may design a gorgeous and flourishing Agapanthus garden that enhances the beauty and joy of your outdoor area every year by establishing specific objectives and taking these important elements into account.

CHAPTER TWO
SETTLING UP

How To Choose The Best Site For Your Agapanthus Garden

For the best possible development and flowering, your agapanthus garden must be placed in the ideal location. Choose a site that receives at least six hours of direct sunlight per day for these gorgeous plants, as they do best in areas with lots of sunshine. They can, however, also withstand some shade, particularly in warmer climates.

Take into account the area's soil drainage when choosing. Agapanthus needs soil that drains properly to avoid waterlogging, which can cause root rot. To increase drainage, you might need to treat your heavy clay soil with organic matter like old manure or compost.

The exposure to the wind should be taken into account. Agapanthus can withstand moderate wind, but too much wind can harm their fragile foliage and blossoms. You can safeguard them by planting them behind taller plants next to a wall, or somewhere that is shielded from high winds.

Finally, consider the location's visual appeal. Agapanthuses are great border plants that provide flower beds' height and color. They also grow well in pots, so if you'd rather container garden or have limited room, think about setting them up in pots on your patio or balcony.

Getting The Soil Ready For Maximum Growth

To give your agapanthus plants the greatest growing conditions possible, you must prepare the soil before planting. First, clear the planting area of any weeds or debris so that your plants have room to grow freely without interference.

Check the pH of the soil to make sure it is between the ideal range of 6.0 and 7.0, which is slightly acidic to neutral for agapanthus. You can add lime to your soil to improve the pH if it's too acidic. On the other hand, if it's excessively alkaline, you can add compost or sulfur to lower the pH.

Next, add organic matter to the soil to enhance its fertility and texture. Peat moss, well-rotted manure, and compost are great options for enhancing soil structure and adding vital nutrients for plant development. Thoroughly incorporate the organic particles into the top 6 to 8 inches of soil.

If you want to give your Agapanthus the nutrients they require for the entire growing season, think about using a slow-release fertilizer designed specifically for flowering plants. For optimal foliage growth at the expense of blossoms, avoid overfertilizing by adhering to the manufacturer's recommended application rates and timing.

When you're ready to plant your agapanthus, make sure the soil is well-watered and uniformly moist but not soggy.

Selecting The Appropriate Agapanthus Types For Your Climate

To guarantee that your agapanthus varieties thrive in your garden, choose the ones that are best suited to your climate. There is bound to be a type of these adaptable plants that meets your demands because they are available in a multitude of shapes, hues, and degrees of cold resistance.

Seek for Agapanthus cultivars labeled as cold hardy or winter tolerant if you reside in a colder climate. These cultivars have a higher chance of surviving and blossoming year after year since they can tolerate below-freezing conditions better.

Select Agapanthus types that can withstand heat and drought in warmer areas. Compared to more delicate types, these can survive in hot, sunny weather and use less water.

Make sure the Agapanthus types you select are the appropriate size for the available space in your yard by taking into account their final size.

While some types can grow several feet tall and wide, others, such as dwarf or compact forms, are better suited for smaller gardens or container planting.

Lastly, consider the color and timing of the blooms on the Agapanthus cultivars you choose. To prolong the display of lovely blossoms in your garden, select colors that go well with the color scheme you currently have in place for your garden and seek kinds that bloom at various periods during the growing season.

CHAPTER THREE

AGAPANTHUS PLANTING

A Comprehensive Guide For Planting Agapanthus Seeds Or Bulbs

The procedure of planting Agapanthus is very simple, regardless of whether you use seeds or bulbs. First, decide on a good spot for your Agapanthus.

These plants like full sun to partial shade and do best in well-draining soil. For best growth, make sure the area gets six hours or more of sunlight each day.

After deciding on the ideal location, get the soil ready. Agapanthus prefers soil with a pH range of 6.0 to 7.0 which is slightly acidic rather than neutral. You can add lime to your soil to improve the pH if it's too acidic. On the other hand, you can add sulfur or peat moss to adjust the pH if it's too alkaline.

When planting Agapanthus bulbs, make sure the hole you dig is deep enough to hold the bulb and roughly double its diameter. Make sure the bulb is planted at the same depth as it was in the pot or nursery container by placing it in the hole with its pointed end pointing up and covering it with soil.

To prepare the soil for Agapanthus seeds, loosen the soil down to a depth of around six inches. Once the soil has been ready, equally distribute the seeds over it and then gently cover them with a thin layer of earth. Remember that Agapanthus seeds need sunlight to sprout, so don't bury them too far down.

Water the soil well after planting to let the dirt settle around the seeds or bulbs. Keep the moisture content constant, particularly during the establishing phase. Once established, agapanthus appreciate slightly dry circumstances, so avoid overwatering them.

Advice On Fertilisation And Watering For Healthful Growth

Your Agapanthus plants need proper watering to stay healthy and vibrant. Water your Agapanthus frequently during the growing season, making sure the soil stays consistently moist but not soggy. To avoid root rot, let the soil somewhat dry out in between waterings.

To reduce the risk of fungal diseases, water straight into the soil rather than overhanging it. To guarantee complete penetration, apply water straight to the root zone using a drip irrigation system or soaker hose.

Fertilization, along with consistent watering, can support robust growth and an abundance of blooms. As soon as new growth starts to appear in the early spring, apply a balanced fertilizer designed for flowering plants. Apply fertilizer at the rates recommended by the manufacturer; overfertilizing might result in excessive growth of foliage at the price of flowers.

To provide a consistent supply of nutrients throughout the growth season, you can also add a liquid fertilizer every four to six weeks. As an alternative, to improve the soil and encourage healthy development, top-dress the area with compost or well-rotted manure.

Mulching Strategies To Preserve Weeds And Preserve Moisture

In your Agapanthus garden, mulching is an easy yet powerful strategy to control soil temperature, retain moisture, and keep weeds at bay. Be careful not to cover the crowns of your plants while applying a two to three-inch layer of organic mulch, such as compost, wood chips, or shredded bark, around the base of your plants.

Mulch acts as a barrier of defense, reducing weed development and evaporation to assist keep soil moisture in place. Furthermore, when organic mulches decompose over time, they improve the structure of the soil and feed it with essential nutrients.

Avoid packing mulch up against the stems of Agapanthus plants while mulching around them, since this might encourage rot and fungal infections. Maintain a modest space to promote airflow and avoid moisture accumulation between the mulch and the base of the plant.

Regularly check your mulch layer and add more as necessary, especially in the event of a significant downpour or windy weather.

An ideal growing environment will be created for your Agapanthus plants by keeping a regular mulch layer; this will ensure that they flourish and yield beautiful blooms every year.

CHAPTER FOUR
TAKING CARE OF AGAPENTHUS
Essential Agapanthus Care Practices

irrigation: Agapanthus plants require proper irrigation to stay healthy. These plants demand soil that is continuously damp but not saturated.

Water deeply once a week during the growing season, which is usually in the spring and summer, or whenever the top inch of soil feels dry to the touch. When the plant is dormant in the autumn and winter, water it less frequently and let the soil partially dry up in between applications.

Soil and Fertilisation: Well-draining soil that is high in organic matter is ideal for agapanthus growth. To enhance drainage, mix compost into compacted or heavy soil. As new growth appears in the spring, fertilize the plants using a balanced, slow-release fertilizer. Steer clear of overfertilizing as this might cause excessive growth of foliage at the price of blooming.

Temperature and Sunlight: Full sun is preferred by these plants over partial shade. Place them in an area where they receive at least six hours of sunlight per day. USDA zones 7 through 11 are generally suitable for agapanthus, although in colder climates plants might need to be protected from frost. During periods of cold, mulching the area around the plant's base can aid in protecting the roots.

Trimming Agapanthus

Deadheading: To promote ongoing flowering and keep the plant from devoting energy to seed production, remove spent flowers regularly. When the blossoms have faded, trim off the flower stems at the base. By keeping the plant neat, deadheading also contributes to its continued aesthetic appeal.

Division: Agapanthus plants may get overcrowded with time, which might result in less vigor and flowering. Every three to five years, divide the plants to encourage better growth and rejuvenation.

Dig up the clumps carefully in late winter or early spring, and remove the individual rhizomes, making sure that each division has many strong roots and shoots. Plant the divisions again, separating them by at least 12 inches in well-prepared soil.

Cleanup: After the foliage turns brown in the autumn, trim it back to the plant's base. This prepares the garden bed for winter and aids in the prevention of fungal illnesses.

To prevent spreading illnesses or pests, dispose of the clipped leaves in a green trash container or compost pile.

Techniques For Controlling Pests And Diseases In Agapanthus

Common Pests: Though they are largely pest-resistant, agapanthus occasionally draws aphids, thrips, or snails.

Keep an eye out for symptoms of infestation, including stunted growth, yellowing leaves, or slime trails, frequently. Use natural predators such as nematodes or predatory insects to manage snail populations, or handpick snails and discard them. As directed by the manufacturer, spray the plants with neem oil or insecticidal soap to get rid of aphids and thrips.

Disease Prevention: In humid environments or soil with poor drainage, agapanthus are particularly vulnerable to fungal diseases such as leaf spot, root rot, and powdery mildew.

Avoid overhead watering, which can encourage the growth of fungi, and make sure there is enough air circulation around the plants by spacing them enough apart to ward off these diseases.

Any unhealthy foliage should be removed and disposed of right once to stop the infection from spreading. Use a fungicide labeled for ornamental plants if necessary, making sure to apply it at the specified rates and times.

Understanding Stress Symptoms And How To Treat Them

Wilting: Underwatering or root rot may be the cause of wilting leaves on your agapanthus plants. Examine the moisture content of the soil and modify your watering schedule accordingly.

If the roots of the damaged plant seem mushy and brown due to waterlogging, carefully take the plant from the soil, remove any decaying roots, and repot it in new soil that drains properly.

Yellowing Leaves: A pest infestation, overwatering, or nutritional shortage may be indicated by yellowing leaves.

To find out whether any vital nutrients are missing from the soil, test the soil and make any necessary amendments.

To avoid soggy circumstances, modify your watering schedule and check your plants for indications of pests like thrips or aphids. To stop additional damage, take quick action to treat any infestations.

Absence of Blooms: If the agapanthus plants on your property do not bloom, it could be because of inadequate sunlight, overcrowding, or excessive fertilizer.

If required, move the plants to a sunnier location to make sure they are getting enough sunshine. Split up dense bunches to encourage improved ventilation and blooming.

Steer clear of overfertilizing as this can cause blossoms to be sacrificed in favor of lush foliage. Instead, to encourage healthy growth and flowering, sparingly use a balanced fertilizer in the spring.

Your agapanthus plants will flourish and provide lovely blossoms to your garden season after season if you adhere to these basic maintenance instructions, pruning methods, and pest and disease control measures.

CHAPTER FIVE

DISSEMBLING AGAPANTHUS

Agapanthus lends a sense of elegance to any garden setting with its eye-catching blue or white blossoms perched on towering stems.

Propagation is the route to take if you want to share the beauty of Agapanthus with loved ones or add to your collection. You can use a variety of approaches, each with its own set of processes and things to think about.

Methods Of Propagation

Split Up

Splitting Agapanthus is arguably the easiest and most popular way to propagate the plant. With this method, the plant is divided into smaller parts, each with its roots and shoots. Select an established Agapanthus plant that has been in the ground for a few years or more to get started. Remove the entire plant with care in the early spring, being cautious not to break the roots. After taking it out of the ground, gently shake off any remaining dirt to reveal the root system.

Next, split the plant into smaller clumps using a clean, sharp knife or garden spade. There should be a few robust-appearing shoots and some root system linked to each clump.

To prevent harming the plant, precise cutting is necessary. After splitting, plant the clumps again, separating them by at least 12 to 18 inches in well-draining soil. After planting, give the area plenty of water to help settle the soil and promote the formation of new roots.

Seeds

Although it can be somewhat more difficult, growing Agapanthus from seed can be quite rewarding. Begin by gathering mature seed pods in late summer or early fall from established Agapanthus plants.

The seeds within should be solid and black, and the pods should normally be dried and brown. Once the seeds are extracted from the pods, gently rub them between your fingers to get rid of any last bits of debris.

It's crucial to carefully prepare the planting space before adding the seeds. Choose a spot that receives lots of sunlight and soil that drains properly.

The seeds can be sown directly in the ground or started indoors in trays or pots. If beginning indoors, add high-quality potting mix to containers and scatter the seeds thickly on the top.

Apply a thin layer of dirt to the seeds and give them a gentle watering.

Reverses

Little plantlets called offsets grow up at the base of established Agapanthus plants. To make fresh individual plants, these can be carefully removed and replanted. Wait until late winter or early spring, when the plants are dormant, to reproduce Agapanthus from offsets.

Make sure each offset has its own set of roots by carefully separating it from the parent plant with a sharp knife or garden spade.

Replant the offsets in containers or straight in the ground after they have been divided. Use soil that drains well, and water your plants well after planting. Until the newly propagated plants grow established, place them in a covered area with indirect sunlight.

Strategies And Timing For Effective Propagation

The success of your endeavors greatly depends on the propagation's timeliness. For division, early spring is best when the plant is just beginning to emerge from dormancy.

It is best to gather and spread seeds in late summer or early autumn so they have time to germinate and establish before winter arrives. When the plant is dormant in late winter or early spring, offsets can be pulled out and replanted.

Whatever approach you decide on, creating the ideal environment is crucial for effective propagation. Planting agapanthus requires well-draining soil and lots of sunlight, so be sure the placement is right. Water newly propagated plants well, but do not soak them too much since this might cause root rot.

Solving Typical Propagation Problems

Although spreading Agapanthus is not too difficult, there are a few frequent problems to be mindful of. One of the most prevalent issues is root rot, which can be brought on by excessive moisture or inadequate drainage in the soil. Planting Agapanthus in well-draining soil and avoiding overwatering are the best ways to stop this.

Poor germination is another thing to be cautious about when propagating from seeds. Before planting, consider soaking the seeds in water for a full day to increase germination rates.

For germination to occur, you should also sow the seeds at the proper depth and supply enough moisture and warmth.

Lastly, it's critical to keep an eye out for any indications of illness or stress in recently propagated plants. Look out for any strange growth patterns, wilting, or yellowing of the leaves. If you find any issues, deal with them right once to save the plant from suffering more harm.

You can effectively reproduce Agapanthus and appreciate the beauty of these magnificent plants in your garden for many years to come by using these suggestions and strategies.

With a little perseverance and attention, you can quickly be rewarded with a flourishing collection of Agapanthus plants, regardless of whether you decide to divide, seed, or grow offsets.

CHAPTER SIX
UTILISING AGAPANTHUS IN DESIGN

Agapanthus can add elegance and beauty to any landscape design with its eye-catching clusters of blue or white flowers atop long, slender stalks.

These gorgeous plants can be used to create borders, focal pieces, or even whole-themed regions in your garden. Let's investigate some imaginative methods to use Agapanthus in design.

Using Agapanthus In Landscape Architecture

Take into account the size, color variations, and growth patterns of agapanthus when incorporating them into your landscape.

Use them as accents along borders or paths, or plant them in clusters to make a statement. Their grass-like foliage contrasts nicely with other plants and adds texture.

Select a color scheme that goes well with the plants you already have for a unified effect.

White agapanthus variants offer a sharp contrast against darker foliage or vivid flowers, while blue varieties go nicely with softer colors like lavender and pastel pink.

To create dynamic compositions, combine Agapanthus with other decorative grasses, shrubs, or perennials.

Try varying the heights and textures of your design to give it more depth and visual appeal.

Ideas For Stunning Combinations Of Companion Plants

Agapanthus grows well with many other companion plants, adding to their beauty and benefiting from each other's growth.

For a low-maintenance, water-wise landscape, combine them with drought-tolerant plants like lavender, rosemary, or decorative grasses.

Combine Agapanthus with lush foliage plants like ferns, hostas, or elephant ears to create a tropical vibe. There will be a rich, exotic ambiance created by the contrasting materials and shapes.

When choosing companions, take seasonal interest into account. Agapanthus can be used with late-blooming perennials like asters or sedums for a colorful autumn show, or it can be used with early-blooming bulbs like tulips or daffodils for a springtime color pop.

Advice For Designing Agapanthus Arrangements To Make A Visual Statement

Agapanthus arrangements can enhance the aesthetic appeal of any area, whether they are placed in flower beds or containers.

To begin, choose the appropriate size and style of container to go with your overall design.

Select permeable pots with adequate drainage to avoid soggy soil, particularly in regions with high rainfall.

Agapanthus variations can be layered for a dramatic effect of height and color. For a graded impression, mix tall and dwarf types; alternatively, combine blue and white blossoms for a dramatic contrast.

To soften the design and add a cascade element, incorporate trailing plants over the sides of containers, such as creeping Jenny or ivy.

If you want to draw pollinators or add seasonal interest, think about including complementary flowers or foliage.

To encourage healthy development and an abundance of blooms, agapanthus should be watered often, especially during hot, dry times, and fertilized with a balanced, slow-release fertilizer.

To ensure that flowers continue to bloom all season long, deadhead spent blossoms.

You can create beautiful landscapes and arrangements that highlight the beauty of agapanthus while improving the overall appeal of your outdoor space by adhering to these design ideas and recommendations.

CHAPTER SEVEN

AGAPANTHUS WITHIN CANDS

Selecting The Proper Soil Mix And Containers For Potted Agapanthus

Growing Agapanthus in pots requires careful selection of the right soil mix and containers. When making your container selection, make sure it has enough space for the plant's root system and growth. Selecting a container that is at least 12 inches deep and wide is a good general rule of thumb.

To avoid waterlogging, which can result in root rot, make sure the containers you've picked have drainage holes at the bottom.

Choose a nutrient-rich, well-draining soil mix that encourages strong root growth and keeps the area from becoming too wet.

To increase drainage, a recommended mixture consists of equal parts garden soil, compost, and coarse sand or perlite. The equilibrium of moisture retention and aeration provided by this mixture promotes the growth of Agapanthus roots.

To further improve drainage, add a layer of gravel or tiny stones to the bottom of pots before planting agapanthus. The plant should then be carefully placed in the center of the container, with the top of the root ball resting just below the rim. Using the prepared soil mix, fill in the remaining area, gently pressing to remove any air pockets.

Tips For Caring For Containers: Watering And Fertilising

Throughout their growing season, Agapanthus plants need to be kept healthy in their containers, which requires proper care. Potted plants are prone to drying out more easily than plants grown in the ground, therefore watering is an essential part of container gardening. Regularly check the moisture content of the soil by sticking your finger up to the first knuckle.

Watering is necessary if the soil seems dry. But keep in mind that overwatering can result in root rot. Instead, water deeply but sparingly to achieve constant moisture levels.

Another crucial component of Agapanthus container maintenance is fertilization. Apply a balanced, slow-releasing fertilizer according to the manufacturer's instructions during the growing season.

This gives the plant the nutrition it needs for strong development and colorful flowers. Furthermore, to encourage continued flowering, think about adding a liquid fertilizer every few weeks as a supplement.

To maintain general plant health and stimulate the creation of new flowers, proper container management also involves deadheading wasted blooms. To avoid disease and keep the foliage looking nice, remove any dead or yellowing leaves.

Overwintering Techniques for Agapanthus Grown in Containers

It's important to overwinter Agapanthus plants cultivated in containers, particularly in areas with harsh winters.

Transfer the containers to a protected area, like a garage or basement, before the first frost to keep them safe from below-freezing conditions. As an alternative, place the containers against a wall that faces south to provide additional warmth or insulate them with bubble wrap.

Watering should be reduced in the winter as the plant needs less moisture because its growth slows down. But watch out that the soil doesn't dry out all the way, as this could harm the roots.

If you live somewhere where it doesn't get very cold, covering the pots with a lot of mulch can act as adequate insulation.

To protect the roots from freezing conditions, add a layer of mulch several inches deep around the base of the plant, covering the soil's surface.

You can guarantee the survival of your container-grown Agapanthus and encourage strong development in the spring by adhering to these overwintering tips.

CHAPTER EIGHT

A AGAPANTHUS FOR SUCCESS IN COMMERCE

Prospects For Agapanthus Growers In The Market

With its eye-catching blossoms and hardiness, agapanthus offers astute gardeners a plethora of commercial prospects.

Success in the agapanthus industry requires a thorough understanding of the market environment. In the landscaping industry, where agapanthus is highly prized for its capacity to infuse color and texture into gardens, public areas, and commercial landscapes, there is a significant possibility for growth.

Agapanthus is frequently used in landscape designs and plans by garden designers, which increases demand for cut flowers as well as potted plants.

The cut flower business is another interesting direction. Agapanthus blooms are a popular choice for bouquets and floral arrangements since they are not only aesthetically pleasing but also have a lengthy vase life.

Growers can profit from the increasing demand for unusual and long-lasting cut flowers by entering this sector and serving florists, event coordinators, and consumers seeking out unusual floral solutions.

Additionally, growers of agapanthus have a chance to market their products as sustainable alternatives in light of the growing number of environmentally concerned consumers.

Once grown, agapanthus is drought-tolerant and requires very little maintenance, making it a green option for both consumers and landscapers. Growers can attract environmentally conscious customers who prioritize sustainability in their purchase decisions by highlighting these eco-friendly traits in their marketing campaigns.

Increasing Production To A Commercial Level

Careful preparation and execution are needed when increasing production to satisfy market demand. One strategy is to grow your enterprise by adding more field or greenhouse production.

This could entail making investments in more machinery, infrastructure, and land to support greater plant volumes.

Incorporating effective production techniques, such as automation and well-designed irrigation systems, can assist increase productivity while lowering labor expenses and resource consumption.

Diversifying your product line to appeal to various market niches is another tactic. Think about branching out into similar products like dried flower arrangements, floral accessories, or even agapanthus-themed merchandise in addition to potted plants and cut flowers. You can reach a larger audience and take advantage of more market opportunities by providing a greater variety of items.

Scaling up an output can also be facilitated by working together with other growers or by establishing alliances with wholesalers, florists, and garden centers. Growers can get economies of scale, open up new distribution channels, and expand their market reach by combining their resources and knowledge. Strategic alliances can further boost competitiveness in the market by offering chances for information exchange, joint marketing initiatives, and research and development.

Sales And Marketing Techniques For Agapanthus Products

Reaching target customers and selling agapanthus items require effective marketing. Using digital channels like email marketing, social media, and e-commerce sites to promote your goods and interact with potential customers is one tactic. Produce eye-catching content that showcases your agapanthus blossoms in a variety of environments, emphasizing their beauty, adaptability, and special features. Promote user-generated content by posting images and client endorsements; engage your audience in conversation to foster brand loyalty and create excitement about your offerings.

Engaging in trade exhibitions, farmers markets, and garden festivals can serve as efficacious marketing tactics to directly connect with consumers. To draw people in, create eye-catching displays of your agapanthus flowers and plants and provide exclusive deals or promotions. Utilize these occasions to build relationships with business associates, network with industry experts, and get client input so that you can keep improving your goods and advertising campaigns.

Additionally, don't discount the effectiveness of conventional marketing strategies like direct mail, signs, and print advertising. Reach out to prospective consumers in your neighborhood by advertising your agapanthus items in local newspapers, gardening publications, and bulletin boards. Make use of captivating pictures and compelling copy to draw viewers in and explain the special advantages of using agapanthus for their floral and gardening needs. You may efficiently increase market recognition, spark interest, and boost sales of your agapanthus items by fusing traditional and digital marketing techniques.

CHAPTER NINE

FAQs AND REGULAR QUESTIONS

Responses To Common Questions Concerning The Cultivation Of Agapanthus

What time of year is ideal for planting Agapanthus? It is ideal to grow agapanthus in the spring when the risk of frost has passed. This enables the plants to take root before the sweltering summer months. In warmer climates, though, you can also plant them in the autumn.

How much sunlight is necessary for Agapanthus plants? Full sun to light shade is ideal for agapanthus growth. For optimum growth and flowering, they must receive at least six hours of sunlight each day. They can, however, withstand some shade, particularly in warmer climates.

Which kind of soil is ideal for Agapanthus plants? Agapanthus likes its soil to be rich in organic content and well-draining. For these plants, a sandy loam or loamy soil is ideal. For best growth, make sure the pH of the soil is slightly acidic to neutral.

How frequently should my Agapanthus be watered? Watering agapanthus plants regularly is necessary, particularly during the growing season. Once or twice a week, give them heavy watering, letting the soil gradually dry out in between. When development slows down in the winter, watering should be reduced.

Do plants that are Agapanthus require fertilization? Fertilizing agapanthus can indeed encourage strong growth and a profusion of flowers. When new growth emerges in the spring, apply a balanced fertilizer designed for flowering plants. Apply again every four to six weeks while the plant is growing.

How are Agapanthus plants divided? Over time, agapanthus can get crowded, which will impair their ability to bloom. Every three to five years, divide well-established clumps to revitalize the plants. Remove the clump by digging it up and carefully split the rhizomes apart, making sure each division has a few strong roots and shoots.

Taking Care Of Typical Problems Like Yellowing Leaves And Insufficient Blooms

Agapanthus leaves that are yellowing can be a sign of several problems, such as pests, nutrient deficits, or overwatering. Evaluate the plant's watering schedule and make any adjustments to solve this issue. Maintain enough drainage to avoid standing water. Additionally, to address nutritional deficits, think about applying a balanced fertilizer. Examine the plant for indications of pests like spider mites or aphids, and apply insecticidal soap if necessary.

Absence of Blooms: There are several reasons why your Agapanthus might not be blooming as it should. Make sure the plant is getting enough sunshine as it is a common cause of this. Divide congested clumps if needed, as overcrowding can also prevent blooming. Check the plant's nutrient levels as well, and fertilize it if needed.

Ultimately, exercise patience, as agapanthus may require a year or two for establishment before bearing a profusion of blossoms.

A Beginners' Guide To Troubleshooting

Wilting Leaves: Wilting leaves may be a sign of root disease or underwatering. Verify the moisture content of the soil and modify your watering schedule as necessary.

Make sure the drainage system is working properly to avoid soggy soil, which can cause root rot. Examine the roots closely for any mushy, discolored spots if you suspect root rot, and cut them as needed.

Stunted Growth: Agapanthus plants may exhibit stunted growth as a result of overcrowding, inadequate sunshine, or poor soil quality. Increase the fertility of your soil by adding organic materials, like compost. To promote greater growth, make sure the plant gets enough sunshine and think about moving clumps that are too close together.

Pest Infestations: Aphids, spider mites, and snails/slugs are among the common pests that harm Agapanthus. Keep an eye out for insect indicators on the plants, including warped leaves or slime trails. Aphids and spider mites can be managed with horticultural oil or insecticidal soap. To protect the plants against snails and slugs, pick them up by hand or use slug pellets. To lessen the habitat of pests, regularly clear weeds and debris from around the plants.

CHAPTER TEN

PROGRESSIVE METHODS AND INNOVATIONS

Sophisticated Cultivation Methods For Skilled Farmers

Expert gardeners of agapanthus frequently look for cutting-edge methods to improve the growth and beauty of these stunning blooms.

Precision watering is one cutting-edge technique that entails giving each plant the precise amount of water it requires depending on its unique requirements. This can be accomplished by employing strategies like moisture sensors to track the moisture content of the soil or drip watering.

Another advanced approach is soil amendment, where gardeners meticulously modify the pH and nutrient levels of the soil to enhance plant health and flowering. This could entail applying compost or other organic waste, or it could entail applying fertilizers specifically designed to meet agapanthus needs.

Furthermore, to encourage healthier development and more plentiful flowering, seasoned growers can experiment with various pruning techniques. This could involve cutting some older leaves to enhance air circulation and lessen the danger of illness, or it could involve deadheading spent blossoms to promote the creation of new flowers.

Advanced growers can also extend the growing season or establish ideal growing conditions for agapanthus by using greenhouse or indoor gardening techniques. This enables people to experiment with uncommon or exotic types that might need more regulated surroundings, or to simply enjoy these beautiful blooms all year round.

Seasoned growers may bring agapanthus plants to their full potential and enjoy even more magnificent flowers year after year by implementing these cutting-edge approaches into their gardening operations.

Advances In Agapanthus Research And Breeding

Discoveries are continually being made in the field of agapanthus breeding and research to enhance plant characteristics including flower color, size, and disease resistance. Breeders can produce new hybrids with distinctive traits by crossing diverse agapanthus kinds, an interesting field of innovation.

To create a hybrid with light blue flowers, for instance, breeders can combine a classic blue agapanthus with a white variety. Alternatively, they might add genes from adjacent plant species to improve characteristics like pest resistance or drought tolerance.

To quickly reproduce desired agapanthus variants, researchers are also investigating methods like tissue culture propagation in addition to hybridization. This enables breeders to ensure uniformity and quality in the market by swiftly producing vast numbers of plants with identical genetic features.

Additionally, improvements in genetic sequencing technology are helping scientists learn more about the genetic composition of agapanthus plants, which will help them understand their development patterns, methods for flowering, and disease resistance. Afterward, new breeding techniques can be created using this knowledge to enhance plant performance as a whole.

All things considered, agapanthus breeding and research continue to yield fascinating new kinds that will delight gardeners everywhere while augmenting the beauty and diversity of these well-liked garden flowers.

Examining Novel Approaches And Opportunities In Agapanthus Horticulture

The world of agapanthus gardening is dynamic, and ever-changing with the arrival of new possibilities and trends with each passing season. Agapanthus, with its dramatic foliage and vivid blooms that can add drama and color to any planting scheme, is a popular choice for mixed borders and container gardens.

The use of agapanthus in sustainable landscaping techniques is another growing trend. These hardy plants are prized for their adaptability to a range of soil conditions and low water requirements. They are therefore the perfect option for environmentally conscientious gardeners who want to design stunning, water-efficient landscapes.

In addition, agapanthus is becoming more and more popular as cut flowers because of its long-lasting blooms, which make them a great option for bouquets and floral arrangements. This creates new opportunities for commercial growers and florists who want to expand their product lines.

Furthermore, protecting rare and endangered agapanthus species in their natural environments is receiving more attention. The goal of botanical gardens and conservation groups is to save these species from unlawful collection and habitat degradation so that future generations can enjoy them.

All things considered, there are plenty of intriguing trends and opportunities in the world of agapanthus gardening for gardeners of all skill levels to discover and appreciate. There's always something new to learn about the world of agapanthus, whether you're a newbie gardener looking to add a splash of color to your yard or an expert grower searching for the newest advancements.